THE HEARTBEAT OF HUMANITY: THE JOURNEY OF RICE

Dr. S Umamaheswari
Dr. Joyce R
Mr. Saktheeswaran C
Ms. Swetha RM
Ms. Priyadarshini
Ms. Aarthi S

TANEESHA PUBLISHING

ISBN : 9788169244275
Edition : First, March 2026
Title : The Heartbeat of Humanity: The Journey of Rice
Author : Dr. S Umamaheswari, Dr. Joyce R, Mr. Saktheeswaran C, Ms. Swetha RM, Ms. Priyadarshini, Ms. Aarthi S

Published by

TANEESHA PUBLISHING

An Imprint of **Taneesha and Prachi NextGen Publication**

Registered Address: Plot No. 48, Khasra No. 646, Road No. 2,
Suman Nagar, Bahadarabad, Haridwar, Uttarakhand – 249402, India
Website : www.taneeshapublishing.com
E-mail : support@tpnextpub.com
Phone : +91 9760418103

Printed by :
Manipal Technologies Limited, Bengaluru - 560001, Karnataka

Index

Sr. No.	Particular	Page No.
1	Introduction	4
2	Objectives of the Study	6
3	Importance of Rice in India	8
4	Land Preparation	13
5	Seed Selection and Nursery for Rice	15
6	Transplantation Process	22
7	Irrigation Methods	25
8	Fertilization and Pest Control	27
9	Harvesting Process	29
10	Threshing, Drying and Milling	35
11	Packaging, Storage and Transportation	38
12	Supply Chain and Economic Importance	40
13	Challenges in Rice Production	44
14	Technology and Future Scope	46
15	Conclusion	48

Introduction

Rice, a humble grain with a global impact, serves as a lifeline for over 3.5 billion people worldwide, providing sustenance and nourishment to more than half of the Earth's population. In countries like India, where rice production is a cornerstone of agriculture, this crop goes beyond mere cultivation—it weaves itself into the fabric of socio-economic existence. From the lush fields of Punjab to the vibrant deltas of Tamil Nadu, the journey of rice embodies a harmonious blend of ancient traditions and modern agricultural practices, shaping rural economies and cultural identities.

The process of bringing rice from the hands of farmers to the plates of consumers is a complex and intricate one, involving meticulous steps such as land preparation, seed selection, transplanting, and post-harvest processing. Each stage requires a delicate balance of labor, resources, and expertise to ensure the production of high-quality grains that meet the demands of a growing population. Along this journey, rice not only sustains livelihoods across the agricultural value chain but also plays a significant role in ensuring food security and economic stability.

Despite its vital importance, the cultivation of rice faces numerous challenges, including climate change, water scarcity, and environmental impacts. However, through innovative practices like Direct Seeded Rice and precision farming, farmers are adapting to these challenges and striving towards a more sustainable future for rice production.

This paper aims to explore the multifaceted journey of rice, shedding light on the pivotal role of farmers, the impact of technological advancements, and the obstacles that must be overcome. By delving into the various stages of rice cultivation, from soil preparation to consumption, this study underscores the essential principles of productivity, equity, and resilience that are crucial for the future of agriculture in India.

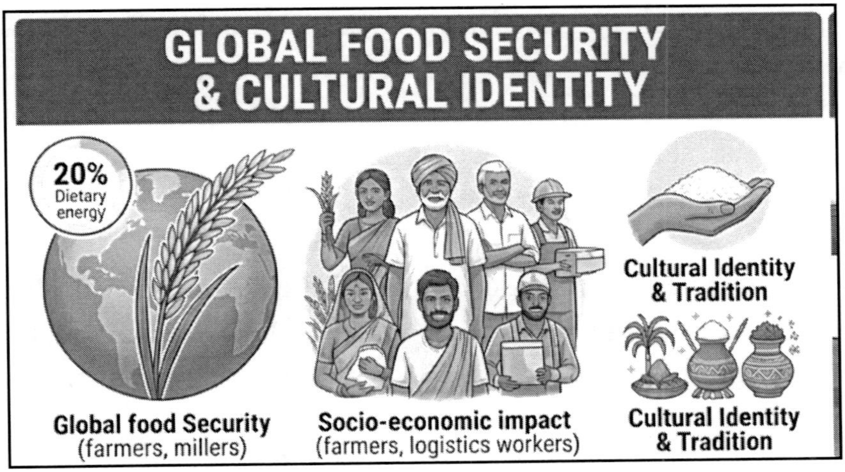

Objectives of the Study

Rice, as India's staple food, demands a complete evaluation of its production adventure to make certain sustainability and financial viability. This take a look at mostly ambitions to map the entire fee chain—from land guidance via consumption—highlighting approaches, inputs, and outputs precise to India's numerous agro-climatic areas just like the Indo-Gangetic plains and southern deltas. A middle objective is to quantify resource efficiencies, along with water utilization per kilogram of rice, focused on reductions from cutting-edge levels of 3,000 liters/kg to beneath 1,500 liters/kg via present day methods.

Secondary desires attention on financial assessment: evaluating farmer incomes, which average ₹ 25,000/ha however may want to upward thrust 50% with mechanization, and supply chain margins wherein millers hold 20-30% cost. The take a look at identifies bottlenecks like put up-harvest losses (10-15%) and proposes interventions, consisting of bloodless garage investments well worth ₹ 50,000 crore wished nationally. It additionally examines social influences, along with employment for 200 million rural employees, with a lens on gender roles wherein ladies handle eighty% of transplanting.

Environmental targets include auditing greenhouse gasoline emissions—rice fields emit 10% of world methane—and recommending practices like trade wetting-drying to reduce this by using 50%. Coverage evaluation covers government schemes like PM-KISAN and national meals security challenge, assessing their role in boosting yields from 2.7 to 4 tons/ha. Academic objectives contain developing farmer schooling modules primarily based on findings, disseminated via 10,000 KVKs.

Technological foresight evaluates drones, AI, and biotech, projecting 20% yield profits via 2030. Facts series methods embody area surveys of 500 farmers across Punjab, Telangana, and Tamil Nadu, GIS mapping, and econometric modeling of charge volatility. Success metrics encompass a 15% improvement in water productivity and coverage briefs for state agriculture departments. Ultimately, this examine bridges academia, industry, and government to support rice as a pillar of India's $500 billion agri-economic system.

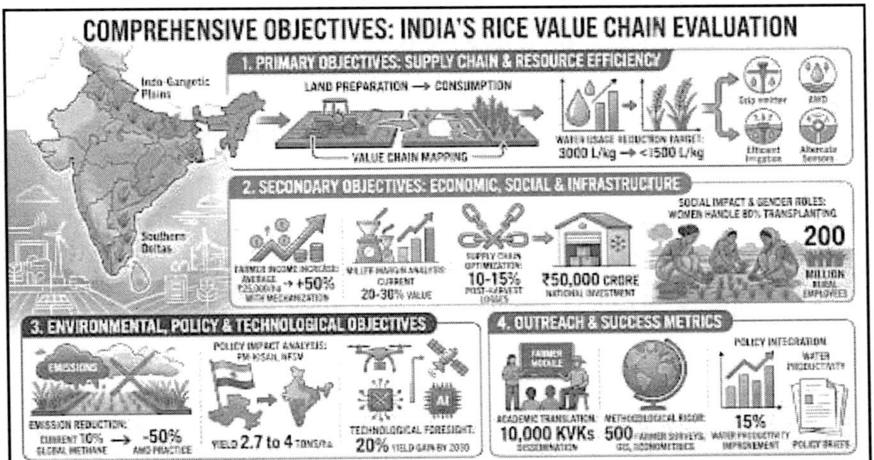

Importance of Rice in India

Rice is one of the most important food crops in India and plays a significant role in the daily lives of millions of people. It is not only a source of food but also an important part of the country's agricultural economy. For centuries, rice has been cultivated in India and has become deeply connected with the nation's culture, traditions, and lifestyle. From rural villages to large cities, rice remains a staple that nourishes people and supports livelihoods. Its cultivation and consumption reflect the agricultural heritage of India.

1. Rice as a Staple Food

Rice is the primary staple food for a large portion of the Indian population, especially in states such as Tamil Nadu, West Bengal, Kerala, Andhra Pradesh, and Assam. It provides essential carbohydrates that give energy for daily activities. Many traditional Indian meals are centered around rice, accompanied by vegetables, lentils, or meat dishes. Because it is easy to cook, digest, and combine with different foods, rice has become an essential part of Indian diets. It is consumed in different forms such as boiled rice, steamed rice, rice flour, and puffed rice.

2. Role of Rice in Daily Meals and Celebrations

Rice is not only a daily food but also an important part of celebrations and special occasions. Many popular Indian dishes are made using rice, such as biryani, pulao, idli, dosa, pongal, and kheer. These dishes are served during festivals, weddings, and family gatherings. Rice-based sweets and snacks are also widely prepared during festive seasons. Because of this, rice holds both nutritional and cultural value in Indian society.

3. Major Rice Producing States in India

India is one of the largest producers of rice in the world. States like West Bengal, Uttar Pradesh, Punjab, Tamil Nadu, Andhra Pradesh, and Odisha contribute significantly to rice production. The warm climate, fertile soil, and adequate water supply in these regions make them suitable for rice cultivation. Monsoon rains play a crucial role in the growth of rice crops. Farmers prepare the land, plant seedlings, manage irrigation, and harvest the crop with great effort to ensure a successful yield.

4. Role of Farmers in Rice Cultivation

Farmers are the backbone of rice production. Rice cultivation requires hard work and careful management throughout the growing season. Farmers prepare the land by ploughing and leveling it, sow seeds in nurseries, and later transplant the seedlings into flooded fields. They also manage irrigation, fertilizers, and pest control to ensure healthy crop growth. After several months of hard work, farmers harvest the mature rice plants, dry them, and send them for processing. Their dedication ensures that rice reaches households across the country.

5. Rice as a Source of Employment

Rice cultivation creates employment opportunities for millions of people in India. Not only farmers but also workers in milling, packaging, transportation, and marketing depend on rice production for their livelihood. Rice mills process the harvested paddy into edible rice, which is then distributed to markets and consumers. Transportation workers help move rice from farms to storage centers and markets. Thus, rice farming supports a large network of jobs and contributes to rural development.

6. Contribution to the Indian Economy

Rice plays a vital role in the Indian economy. India is one of the leading exporters of rice, especially Basmati rice, which is highly valued worldwide for its aroma and long grains. Many countries import Indian rice, generating significant foreign exchange for the nation. The rice industry also contributes to agricultural income and strengthens the country's food supply chain.

7. Rice and Food Security

Rice is essential for maintaining food security in India. The government ensures that rice reaches people at affordable prices through various welfare programs. The Public Distribution System (PDS) distributes rice and other essential food grains to economically weaker sections of society at subsidized rates. This helps reduce hunger and ensures that even low-income families can access basic food supplies.

8. Role of the Food Corporation of India (FCI)

The Food Corporation of India (FCI) plays an important role in managing the supply of rice. It procures rice from farmers at a minimum support price (MSP), stores it in warehouses, and distributes it to different states. This system helps stabilize prices, supports farmers, and ensures that the country has enough food reserves during times of shortage or emergencies.

9. Cultural and Religious Significance of Rice

Rice has deep cultural and religious importance in India. It is commonly used in religious rituals, temple offerings, and traditional ceremonies. In many Indian weddings and festivals, rice is used as a symbol of prosperity and blessings. Throwing rice during ceremonies often represents good fortune and fertility. Because of these traditions, rice holds a special spiritual and cultural value in Indian society.

10. Diversity of Rice Varieties and Dishes

India has a wide variety of rice types, including Basmati, Sona Masuri, Ponni, and Jasmine rice. Each variety has its own taste, texture, and cooking style. Different regions of India prepare unique rice-based dishes according to their cultural preferences. For example, biryani is popular in Hyderabad, pongal in Tamil Nadu, and pulao in North India. This diversity reflects the rich culinary heritage of the country.

11. Conclusion

In conclusion, rice is much more than just a food grain in India. It is deeply connected to the country's agriculture, economy, culture, and traditions. It provides nutrition to millions of people, supports the livelihoods of farmers and workers, and plays an important role in festivals and rituals. As India continues to grow and develop, rice will remain a vital part of the nation's food system and cultural identity.

Agriculture and Economy
National Agricultural Backbone | Millions Employed
Major Producing State: Tamil Nadu
Sona Masuri & Ponni
Farmers' Role and Variety
Basmati
Basmati
Local Varieties
The Staple & Diversity of Dishes
in Tamil Nadu
Biryani
(Hyderabad)
Pongal
(Tamil Nadu)
Idli & Dosa
(South India)
Steamed Rice
Celebrations and Festivals (Cultural Value)
Ponga-o-Pongal!
Food Security & PDS
FCI
FCI
Public Distribution System - Ensuring Food for All
Contribution to Economy & Trade (The connection)
Leading Global Rice Exporter
Storage and Distribution
FCI Warehousing | Stable Food Reserve

Land Preparation

Land preparation plays a crucial role in rice cultivation as it creates the optimal soil environment for rice plant growth. Proper land preparation helps in enhancing soil fertility, managing weeds, ensuring proper water distribution, and supporting healthy root development. Since rice is predominantly grown in waterlogged conditions, it is essential to prepare the field carefully before planting.

The first step in land preparation involves clearing the field of weeds, stones, previous crop residues, and other unwanted materials. This helps prevent pest infestations and ensures that rice plants can grow without competition for nutrients. The next step is ploughing or tilling the soil using tractors, power tillers, or traditional animal-drawn ploughs. Ploughing helps loosen the soil, break hard layers, improve soil aeration, and incorporate organic matter and nutrients into the soil. The field is usually ploughed multiple times to make the soil soft and suitable for rice cultivation.

After ploughing, the field is leveled to ensure even water distribution across the field. Proper leveling is crucial for rice farming as uneven land can lead to waterlogging in some areas and water deficiency in others, affecting plant growth and yield.

Puddling is another important process where water is added to the field and the soil is mixed thoroughly to create a soft, muddy surface. Puddling reduces water seepage, facilitates easy transplanting of rice seedlings, and helps control weeds by burying them under the soil. Farmers also construct bunds around the field to retain water and prevent overflow. Effective water management is essential as rice plants require standing water during their initial growth stages.

During land preparation, farmers may apply organic manure, compost, or farmyard manure to enhance soil fertility. Additionally, basic fertilizers containing essential nutrients like nitrogen, phosphorus, and potassium may be added to improve crop growth and productivity. Modern farming techniques such as laser land leveling and mechanized ploughing can save time, labor, and water while increasing crop yield.

In conclusion, proper land preparation is vital for successful rice cultivation. It ensures good soil structure, efficient water management, reduced weed growth, and improved nutrient availability, ultimately leading to healthy rice plants and higher agricultural productivity.

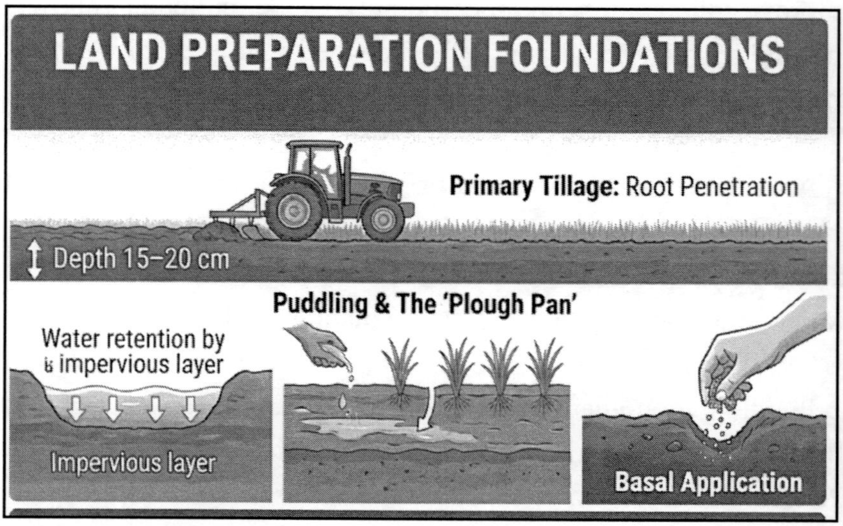

Seed Selection and Nursery for Rice

1. Introduction

Rice, a staple food for millions worldwide, holds a significant place in agriculture, especially in countries like India. The success of rice cultivation hinges on the quality of seeds and the meticulous preparation of the nursery. Healthy seeds and well-managed nurseries are the bedrock of robust seedlings, optimal plant growth, and increased yield, making seed selection and nursery management pivotal in successful rice production.

2. Seed Selection

Seed selection, a critical process in agriculture, involves choosing superior-quality seeds for cultivation. Farmers meticulously handpick seeds to ensure robust crop growth and enhanced productivity. High-quality seeds foster plant strength, disease resistance, and bountiful grain production.

Significance of Seed Selection

Seed selection is a cornerstone of rice cultivation, directly impacting germination rates, plant vigor, and final yield. Inferior seeds can lead to uneven crop growth, reduced grain output, and heightened vulnerability to pests and diseases.

Qualities of Superior Seeds

Farmers should opt for seeds exhibiting the following characteristics:

1. High Germination Capacity

Seeds with a high germination rate ensure uniform and swift sprouting, fostering healthy seedling growth.

2. **Disease Resistance**

Selecting seeds resistant to prevalent rice diseases like blast disease and bacterial leaf blight is crucial for crop protection.

3. **High Yield Potential**

Farmers often prefer high-yielding rice varieties to boost production levels.

4. **Uniform Size and Shape**

Healthy seeds typically boast uniform size and shape, while damaged or shriveled seeds should be discarded.

5. **Seed Purity**

Seeds must be free from impurities like weeds, stones, and dust to ensure optimal growth.

6. **Adaptability to Local Climate**

Choosing seeds suited to the local climate, soil type, and water availability is essential for successful cultivation.

Methods of Seed Selection

Farmers employ various techniques to identify the best seeds for planting.

1. **Hand Selection**

In traditional farming practices, farmers manually sift through seeds, discarding damaged or unhealthy grains. Though simple, this method demands keen observation.

2. **Salt Water Test**

A prevalent method among farmers, the salt water test aids in identifying healthy seeds.

Procedure:

1. Fill a bucket with water.
2. Add salt to increase water density.
3. Pour seeds into the salt water solution.

4. Healthy seeds sink, while damaged ones float and can be removed.

This method aids farmers in selecting robust and viable seeds for cultivation.

3. Seed Treatment

Post-selection, seeds undergo treatment to shield them from diseases and pests. Treatment methods may involve soaking seeds in water or utilizing chemical or biological agents.

Common treatments include:

Soaking seeds in clean water for 12–24 hours

Applying fungicides to prevent fungal infections

Using biofertilizers to enhance seedling growth

Seed treatment enhances germination rates and safeguards young plants during initial growth stages.

4. Nursery Preparation

A nursery serves as a specialized area where rice seeds are sown and nurtured into seedlings before transplantation to the main field. Nurseries offer a controlled environment for young plants to flourish healthily.

Nursery preparation stands as a crucial phase in rice farming, as robust seedlings pave the way for optimal crop establishment and heightened yields.

Types of Rice Nurseries

Various types of nurseries are employed in rice cultivation based on regional and farming preferences.

1. **Wet Bed Nursery**

Common in irrigated rice farming, this method involves preparing nursery beds in moist soil and maintaining adequate water levels.

2. **Dry Bed Nursery**

In this method, seeds are sown on dry soil beds, with water added post-germination.

3. **Dapog Nursery**

Primarily used in certain Asian countries, this method entails growing seeds on a thin layer of soil or organic material for rapid seedling production.

Steps in Nursery Preparation

Effective nursery preparation entails several crucial steps.

1. **Selection of Nursery Site**

Choose a nursery site with care, ensuring it:

Is near a water source

Boasts fertile soil

Has good drainage

Is in proximity to the main field for ease of transportation

2. **Land Preparation**

Plough and level the nursery land, ensuring soft, well-pulverized soil for optimal seed germination.

Steps include:

Ploughing to loosen soil

Weed and stone removal

Land leveling for even water distribution

Division into small nursery beds

3. **Bed Formation**

Prepare small, slightly raised beds in the nursery area to prevent water stagnation.

Adequate spacing between beds facilitates irrigation and drainage.

4. Seed Soaking and Pre-Germination

Soak seeds in water for 12–24 hours before sowing. Allow seeds to sprout slightly in a warm location for another 24 hours, a process known as pre-germination.

Pre-germinated seeds exhibit faster growth and uniform seedling development.

5. Sowing Seeds

Evenly spread sprouted seeds over the prepared nursery beds for uniform distribution, reducing nutrient competition among seedlings.

Follow recommended seed rates to prevent overcrowding.

6. Water Management

Maintain optimal soil moisture during germination, avoiding excess water to prevent seed rot.

Regular light irrigation supports seedling growth.

7. Fertilizer Application

Apply minimal amounts of fertilizer to nursery beds to enhance seedling health. Organic or chemical fertilizers may be used based on farming practices.

8. Weed and Pest Control

Weed out unwanted plants and employ safe pest control measures to protect young rice seedlings.

4. Growth of Rice Seedlings

Rice seedlings typically spend 20–25 days in the nursery, developing robust roots and leaves. Careful nurturing during this phase ensures healthy plant growth.

When seedlings reach 15–20 cm in height, they are ready for

transplantation to the main field.

5. Transplantation Preparation

Before transplantation, water nursery beds to facilitate seedling removal without root damage. Uproot seedlings with care, bundling them for transport to the main field.

Healthy seedlings are then transplanted into the prepared paddy field, where they continue to thrive until harvest.

6. Advantages of Proper Seed Selection and Nursery Management

Effective seed selection and nursery preparation offer numerous benefits:

Promote healthy, robust seedlings

Enhance crop uniformity

Reduce pest and disease risks

Boost rice yield

Facilitate superior plant growth and development

7. Conclusion

Seed selection and nursery preparation stand as pivotal stages in rice cultivation. Opting for high-quality seeds ensures robust germination and plant vigor, while meticulous nursery management provides an ideal setting for seedling growth. By adhering to these practices diligently, farmers can achieve heightened productivity and superior rice quality. Thus, seed selection and nursery preparation play a vital role in the successful cultivation of rice.

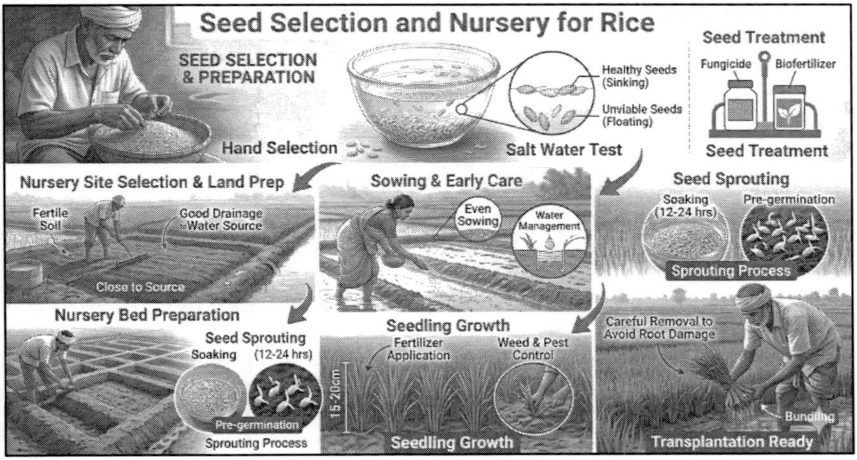
Seed Selection and Nursery for Rice
SEED SELECTION & PREPARATION
Hand Selection
Healthy Seeds (Sinking)
Unviable Seeds (Floating)
Salt Water Test
Seed Treatment
Fungicide
Biofertilizer
Seed Treatment
Nursery Site Selection & Land Prep
Fertile Soil
Good Drainage to Water Source
Close to Source
Sowing & Early Care
Even Sowing
Water Management
Seed Sprouting
Soaking (12-24 hrs)
Pre-germination
Sprouting Process
Nursery Bed Preparation
Seed Sprouting
Soaking (12-24 hrs)
Pre-germination
Sprouting Process
Seedling Growth
Fertilizer Application
Weed & Pest Control
15-20cm
Seedling Growth
Careful Removal to Avoid Root Damage
Bundling
Transplantation Ready

Transplantation Process

The Journey of Rice Planting

Growing rice through the transplantation process is a common practice among farmers in India. This method involves nurturing rice seedlings in a nursery before carefully moving them to the main field. The process not only promotes robust plant growth but also maximizes seed efficiency, resulting in higher crop yields.

1. Cultivating the Nursery

The initial step in the transplantation process is establishing a nursery bed. Farmers meticulously select a small plot of land, diligently ploughing and leveling the soil. Premium quality seeds are soaked in water, sometimes treated with fertilizers or protective chemicals. These treated seeds are then delicately sown in the nursery bed, where they sprout into young rice seedlings within 20-30 days.

2. Preparing the Main Field

While the seedlings flourish in the nursery, farmers diligently prepare the main field. The land is thoroughly ploughed, and water is added to create a soft, muddy texture through a process known as puddling. Puddling aids in water retention and facilitates the seamless transplantation of the seedlings.

3. Upbringing the Seedlings

Once the seedlings reach a height of approximately 15-20 cm after 3-4 weeks, farmers gently uproot them from the nursery, ensuring the roots remain unharmed. The seedlings are often bundled together for easy transportation to the main field.

4. Planting the Seedlings

During this phase, the seedlings are carefully planted in the prepared muddy field. Farmers strategically place the seedlings in rows with adequate spacing between them, typically planting 2-3 seedlings together at each spot. Proper spacing guarantees each plant receives ample sunlight, nutrients, and water for optimal growth.

5. Water Management

Maintaining the correct water level in rice fields during the initial growth stages is crucial. Farmers regulate the water to suppress weeds and support plant development. However, the water level is adjusted or drained at specific growth stages of the crop.

6. Post-Transplantation Care

Following transplantation, farmers implement various measures to promote healthy crop growth. These include fertilization, weed removal, and safeguarding plants against pests and diseases. With attentive care, rice plants thrive and yield a bountiful harvest.

Conclusion

The transplantation process plays a vital role in rice cultivation, enabling farmers to nurture robust plants and achieve optimal yields. Despite the labor-intensive nature of transplantation, it ensures proper plant spacing, effective crop management, and enhanced rice production.

Irrigation Methods

Irrigation is the lifeline of rice, which requires 1, two hundred-1,500 mm of water in step with season, more than some other crop. Conventional flood or continuous submergence approach floods fields to 5-10 cm depth from transplanting to flowering, sourced thru canals (fortytwo million ha irrigated), tube wells (25 million ha), and tanks. In Punjab-Haryana, this helps 7-eight lots/ha yields but depletes groundwater at 0.8 m/yr, with 80% blocks overexploited.

Drip irrigation revolutionizes this by emitters located 30 cm apart turning in 2-4 liters/hour at once to roots, achieving eighty-90% performance versus flood's 40%. Adopted on zero. Five million ha, it saves 50-60% water (500-seven-hundred mm/season), raises yields 15-25% through particular scheduling, and cuts strength prices 30% as pumps run intermittently. In Maharashtra's command regions, hybrids like MTU-7029 below drip yield 9 tons/ha; government subsidies cover 70% set up prices.

Trade Wetting and Drying (AWD), promoted with the aid of IRRI, lets soil dry to 15 cm below surface before reflooding, saving 25-35% water and lowering methane through 48% without yield loss. Subject trials in Odisha display 6. five heaps/ha as opposed to 6 lots in flood; secure AWD thresholds use perforated pipes as indicators. Machine of Rice Intensification (SRI) combines shallow irrigation (2-five cm), unmarried seedling hills spaced 25x25 cm, and organic compost, the use of 30-50% less water for 20-50% higher yields (eight-12 lots/ha in Tamil Nadu).

Sprinkler structures, low-stress rain guns masking 1 ha/set, healthy upland direct-seeded rice (DSR), saving 40% water and permitting 0-until with satisfied seeders to retain stubble. DSR, sown like wheat via seed drills, avoids puddling, cuts methane 70%, and exertions forty%; 2 million ha in Haryana. Precision techniques like tensiometers degree soil moisture for app-primarily based scheduling (e.g., Rice health practitioner app).

Subsurface drip buries tubes 15 cm deep, stopping clogging from algae. Canal modernization with SCADA automates gates for equitable distribution. Challenges encompass preliminary costs (₹ eighty,000/ha for drip) and salinity in coastal Andhra; solutions through micro-irrigation finances (₹ five,000 crore allocated). Future integration with IoT sensors forecasting and many others (evapotranspiration) could optimize to ten kg grain/mm water, as opposed to modern zero.

Rice fields prepared with drip irrigation traces and manage valves show water-saving era in movement.

Fertilization and Pest Control

Ensuring the health and vitality of rice plants is essential in cultivation, with fertilization and pest control playing key roles in this process. Fertilization involves providing necessary nutrients like Nitrogen, Phosphorus, and Potassium to the soil to support plant growth. Nitrogen helps with leaf development, Phosphorus strengthens roots, and Potassium boosts disease resistance. Farmers use a range of fertilizers, including urea, DAP, and potash, at different growth stages to ensure plants receive the nutrients they need. Organic fertilizers like farmyard manure and compost are also utilized to enhance soil fertility and promote sustainable farming practices.

Pest control is crucial as well, as pests such as stem borers, leaf folders, and rice bugs can harm crops and reduce yields. Farmers employ various methods like chemical, biological, and cultural controls to protect their crops. Integrated Pest Management (IPM) is a holistic approach that combines different control methods in an eco-friendly way. By implementing proper fertilization techniques and effective pest control measures, farmers can boost rice productivity, enhance grain quality, and secure a reliable food source for the community.

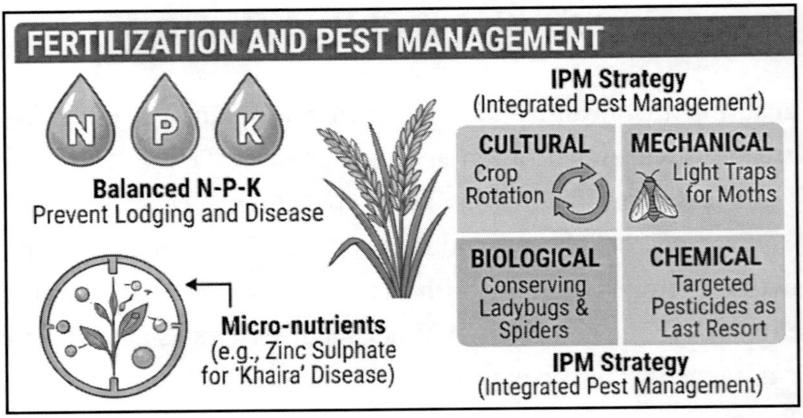
FERTILIZATION AND PEST MANAGEMENT
N
P
K
Balanced N-P-K
Prevent Lodging and Disease
Micro-nutrients
(e.g., Zinc Sulphate for 'Khaira' Disease)
IPM Strategy
(Integrated Pest Management)
CULTURAL
Crop Rotation
MECHANICAL
Light Traps for Moths
BIOLOGICAL
Conserving Ladybugs & Spiders
CHEMICAL
Targeted Pesticides as Last Resort
IPM Strategy
(Integrated Pest Management)

Harvesting Process

Harvesting Rice: A Humanized Approach

Introduction:

The process of harvesting rice marks the culmination of hard work and dedication put in by farmers throughout the cultivation cycle. It is a crucial stage that not only signifies the end of the growing season but also holds the key to maximizing yield and maintaining the quality of the rice grains. Proper harvesting practices are essential to ensure that the grains are mature and of optimal weight. Harvesting too early can result in immature grains, while delaying the process may lead to grain loss due to pests, birds, or adverse weather conditions.

Rice harvesting goes beyond simply cutting the crop; it encompasses a series of post-harvest operations such as threshing, drying, cleaning, and storing the grains. Each of these steps plays a vital role in safeguarding the rice from damage and preserving its quality for consumption and sale.

While traditional methods are still prevalent in many parts of the world, modern agricultural technology has introduced machinery that streamlines the harvesting process, making it faster and more efficient.

Maturity of Rice Crop:

Before embarking on the harvesting process, farmers must assess the maturity of the rice crop to ensure optimal quality and yield. Typically, rice crops mature around 30 to 40 days after flowering, with variations depending on the rice variety and environmental factors.

Signs of a Mature Rice Crop:

Farmers rely on several visual cues to determine the maturity of the rice crop:

1. **Color Change:**

The transition of rice plants from green to a golden yellow hue indicates maturity.

2. **Hardness of Grains:**

Mature grains exhibit firmness when pressed between the fingers.

3. **Drying of Leaves and Stems:**

The drying and loss of green color in the leaves and stems signify readiness for harvesting.

4. **Drooping Panicles:**

The bending of grain heads, known as panicles, due to the weight of mature grains.

5. **Low Moisture Content:**

A reduction in moisture content to approximately 20–25% signals that the crop is ripe for harvesting.

Careful observation of these indicators enables farmers to pinpoint the optimal time for harvesting, ensuring high-quality grains.

Methods of Harvesting Rice:

Two primary methods are employed for harvesting rice: manual harvesting and mechanical harvesting.

Manual Harvesting:

Manual harvesting, a traditional approach prevalent in small farms and rural areas, involves the use of basic tools such as sickles to cut the rice plants. While labor-intensive and time-consuming manual harvesting remains popular due to its simplicity and cost-effectiveness.

Steps in Manual Harvesting:

1. Farmers grasp a bunch of rice plants with one hand.

2. Using a sickle, they cut the plants close to the ground.

3. Harvested plants are gathered and tied into small bundles.

4. Bundles are left in the field for drying.

Mechanical Harvesting:

In contrast, large farms and modern agricultural systems utilize machines like combine harvesters for efficient harvesting. These machines perform multiple operations simultaneously, including cutting the rice plants, threshing the grains, and cleaning them. Mechanical harvesting significantly reduces labor costs, minimizes grain loss, and enhances efficiency in large-scale rice production.

Threshing:

Following the harvesting of rice plants, the next step is threshing, which involves separating the rice grains from the stalks and husks.

Traditional Threshing Methods:

1. Beating Method:

Farmers beat harvested bundles against a hard surface to dislodge the grains.

2. Animal Trampling:

In some rural areas, animals are used to walk over the harvested plants to facilitate grain separation.

Mechanical Threshing:

Modern farmers opt for threshers, machines that expedite the separation of grains from stalks, saving time and labor. Threshing is a critical stage as it ensures efficient collection of rice grains.

Drying of Rice Grains:

After threshing, rice grains retain moisture that must be removed to prevent fungal growth, spoilage, and damage during storage. Drying is a vital component of the harvesting process.

Drying Methods:

1. **Sun Drying:**

Farmers commonly employ sun drying by spreading grains on mats, tarpaulins, or cement floors for exposure to sunlight.

2. **Mechanical Drying:**

Large farms and rice mills utilize mechanical dryers for rapid and efficient moisture removal.

The ideal moisture content for storing rice grains ranges from 12–14%.

Cleaning and Winnowing:

Following drying, rice grains undergo cleaning to eliminate impurities such as straw, dust, husk, and broken grains.

Winnowing Process:

Winnowing, a traditional technique, involves separating lighter materials from heavier grains. Farmers toss grains into the air using a basket or tray, allowing the wind to blow away lighter particles while the heavier rice grains settle back into the container. Modern agriculture employs cleaning machines for enhanced efficiency in this process.

Storage of Rice:

Proper storage is the final step post-harvest, safeguarding rice grains from insects, moisture, rodents, and spoilage.

Common Storage Methods:

1. **Storage Bags:**

Rice grains are packed in jute or plastic bags and stored in warehouses.

2. **Granaries:**

Traditional storage in wooden, clay, or metal granaries.

3. **Modern Storage Silos:**

Metal silos are utilized by large-scale farmers and industries for safe storage of significant rice quantities.

Effective storage ensures the longevity and quality of rice for consumption.

Importance of Proper Harvesting:

Adhering to proper harvesting practices yields numerous benefits:

- Reduces grain loss
- Maintains grain quality
- Enhances market value of rice
- Prevents pest and disease damage
- Increases farmers' income
- Ensures a stable food supply

Careful harvesting and post-harvest handling are imperative for successful rice production.

Challenges in Rice Harvesting:

Farmers encounter various challenges during the harvesting process, including:

- Labor shortages
- Unfavorable weather conditions
- Pest and bird infestations
- Lack of modern machinery
- High equipment costs

The adoption of modern agricultural technologies and improved farming methods helps mitigate these challenges.

Conclusion:

Harvesting rice is a pivotal stage in cultivation that dictates the final yield and quality of the crop. It involves a series of meticulous steps, from cutting mature rice plants to drying, cleaning, and storing the grains effectively. Harvesting at the right time and employing proper post-harvest techniques are essential for minimizing losses and preserving rice quality. While modern machinery has revolutionized the harvesting process, traditional methods remain prevalent in rural areas, underscoring the rich diversity in agricultural practices.

Threshing, Drying and Milling

The Human Side of Rice Processing

Once the rice crop is harvested, a series of post-harvest activities are undertaken to prepare the rice for storage and consumption. These crucial steps involve threshing, drying, and milling, all aimed at transforming the raw rice into a form that is ready to be enjoyed.

1. Threshing

Threshing marks the beginning of the journey to separate the rice grains from the harvested stalks. Farmers gather the rice plants into bundles and bring them to designated threshing areas.

Various methods are employed for threshing:

- **Hands-on approach:** Farmers manually beat the bundles against hard surfaces or wooden platforms to release the grains.
- **Animal assistance:** In some regions, animals are utilized to walk over the harvested crop, aiding in loosening the grains.
- **Mechanical marvels:** Modern farms utilize threshers, machines designed to efficiently separate the grains from the stalks.

Once threshing is complete, the grains obtained are known as paddy, still encased in husk and retaining some moisture.

2. Drying

Drying plays a crucial role in reducing the moisture content of the paddy. Freshly harvested rice tends to have high moisture levels, which, if left unchecked, can lead to spoilage or fungal growth during storage.

Common drying methods include:

- **Sun-powered:** Paddy grains are spread out on mats, floors, or drying yards, basking in the sun for hours or days. Farmers diligently turn the grains to ensure uniform drying.

- **Mechanical assistance:** Large-scale farming operations employ specialized drying machines to expedite the drying process and ensure uniformity.

Proper drying not only safeguards the quality of the rice but also protects it from damage during storage.

3. Milling

Milling is the transformative process of shedding the outer layers of the paddy grain to reveal the edible rice within. This intricate process is typically carried out in rice mills using advanced machinery.

Key steps in the milling process include:

1. **Purification:** Eliminating dust, stones, and other impurities.
2. **Husk removal:** Stripping away the outer husk layer from the paddy grain.
3. **Shining bright:** Polishing the rice to achieve the desired white appearance.
4. **Quality check:** Sorting the rice grains based on size and quality

Upon completion of milling, the rice is now primed for cooking and consumption.

Conclusion

Threshing, drying, and milling stand as vital post-harvest procedures in the realm of rice production. These processes not only serve to separate the grains and reduce moisture but also play a pivotal role in transforming paddy into pristine, edible rice. By handling these steps with care and precision, farmers ensure the delivery of high-quality rice while minimizing losses during storage

Packaging, Storage and Transportation

Post-milling, rice first-class hinges on packaging to maintain 14% moisture, aroma, and wholesomeness. primary packaging makes use of BOPP laminated PP baggage (5-50 kg) with five-layer extrusion for oxygen/moisture limitations (WVTR <5 $g/m^2/day$), revealed thru flexography for branding. Retail packs (1-five kg) rent stand-up pouches with zippers or nitrogen flushing (ninetynine.5% N_2) extending shelf existence to 24 months by way of inhibiting lipid oxidation.

Jute bags (50-100 kg), f6ba901c5019ebe39975adc2eb223bef but absorbent, line with LDPE for bulk FCI procurement (forty million tons/yr). Vacuum packaging eliminates 90% air, preventing weevil infestation; retort pouches for cooked rice withstand sterilization. Biodegradable options like PLA from corn starch emerge, although value 20% better.

Garage infrastructure totals 85 million lots capability: 70% included yards/godowns, 20% silos (15 million tons), 10% MAFW (modified environment flat warehouses) purging CO_2 . FCI's 18,000 godowns keep 12-14% moisture via dehumidifiers (RH 60-sixtyfive%); phosphine drugs fumigate quarterly, killing 100% pests in 7 days. Pusa Bin hermetic baggage (60 kg) allow 0-chemical storage for small farmers, retaining germination 98% for three hundred and sixtyfive days.

Logistics optimization uses RFID/GPS for real-time monitoring reducing pilferage five%; multi-modal corridors like jap Freight corridor slash Delhi-Chennai time from 7 to 4 days. submit-harvest losses drop from 12% to six% with vibration-dampened stacking and humidity loggers. economic scale: deliver chain adds ₹ 20/kg price, producing 15 million jobs; blockchain pilots hint 10,000 lots from farm to fork.

challenges: monsoon delays spike 10% losses; answers consist of drone shipping trials and EV fleets below repute scheme.

Rows of packaged rice baggage loaded onto vans and stored in contemporary silos illustrate seamless logistics.

Supply Chain and Economic Importance

1. The Heart of Rice: Understanding the Supply Chain and Economic Significance

Rice stands as a cornerstone in the global food landscape, serving as a staple for over half of the world's population. Its significance is deeply rooted in the daily diets of millions, particularly in countries like India, China, Bangladesh, Thailand, and Vietnam. In India, rice not only sustains as a primary food source but also stands as a vital agricultural commodity that upholds the livelihoods of countless farmers and laborers.

The journey of rice, from the nurturing hands of farmers to the plates of consumers, unfolds through a series of interconnected stages known as the rice supply chain. This intricate web encompasses production, harvesting, processing, storage, transportation, distribution, and marketing, with diverse individuals and organizations collaborating to ensure the delivery of quality rice to consumers.

Beyond its role as a dietary staple, rice holds immense economic importance. It serves as a significant contributor to employment, agricultural income, rural development, and international trade, supporting a wide array of individuals from farmers and mill workers to traders and retailers, thereby bolstering the economy and fostering growth.

2. Nurturing the Rice Supply Chain: A Closer Look at the Journey

The rice supply chain encapsulates the comprehensive process through which rice traverses from the fields to the hands of consumers, involving numerous stages and participants dedicated to the production and dissemination of rice.

Key stages in the rice supply chain include:

1. Cultivation
2. Harvesting
3. Processing
4. Storage
5. Transportation
6. Distribution and Marketing
7. Consumption

3. **Cultivating the Essence:** The Art of Rice Production

The inception of the rice supply chain commences with the cultivation of rice in agricultural fields. Farmers meticulously prepare the land through ploughing and leveling, followed by the selection and sowing of high-quality seeds in nurseries. The seedlings are then transplanted into the main paddy field, where farmers provide essential irrigation, fertilizers, and pest control measures to nurture healthy plant growth. The growth cycle of rice plants typically spans 3 to 6 months, contingent upon the variety and climatic conditions. Once the crop reaches maturity, it is primed for harvesting.

Rice production stands as a pivotal stage in the supply chain, dictating the quantity and quality of rice available for consumption.

4. **Harvesting and Beyond:** Post-Harvest Endeavors

Upon the maturation of the rice crop, farmers embark on the harvesting process, utilizing either manual methods with sickles or mechanized approaches with combine harvesters to ensure timely and quality harvests. Post-harvest activities ensue, encompassing essential processes such as:

Threshing: Separating rice grains from stalks manually or with threshing machines.

Drying: Reducing moisture content through sunlight exposure or mechanical drying to prevent spoilage.

Cleaning: Removing impurities like dust, stones, and other contaminants to prepare rice grains for processing and storage.

These activities lay the groundwork for subsequent processing and storage of rice.

5. **Refinement in Rice Mills:** Elevating Rice Processing

Following harvesting and cleaning, the paddy undergoes transportation to rice mills for processing, a crucial step in converting paddy into consumable rice. The processing journey entails key steps such as:

1. **Cleaning:** Eliminating dust, stones, and unwanted materials.
2. **Husking:** Removing the outer husk layer from the grain.
3. **Milling:** Extracting the bran layer to yield white rice.
4. **Polishing:** Enhancing the appearance and texture of rice.
5. **Grading:** Sorting rice based on size and quality.

Rice mills play a pivotal role in upholding the quality and market value of rice.

The Heart of Rice:
Understanding the Supply Chain and Economic Significance
1 CULTIVATION: LAND PREP & PLANTING
Ploughing & Leveling
Nursery & Seed Sowing
Selected Quality Seeds
Seedling Nursery
Transplantation
2 The Rice Supply Chain: Field to Plate
Irrigation
Growth Factors
Fertilizers
Pest Control
Integrated pest management
Helpful
Health
Helpful practices
Safe practices
Growth Cycle
Fertilizers
Small → (3-6 MONTHS) → Maturity indicators
2 HARVESTING & POST-HARVEST
Manual Harvesting
Mechanical Harvesting
Post-Harvest Endeavors
Threshing
Grain Separation
Drying
Cleaning
Remove impurities
Timely and Quality Harvests
3 REFINEMENT IN RICE MILLS
Paddy from Fields
Cleaning
Cleaning
Husking
Milling
Polishing
Grading
Rice mills maintain quality and market value.
Milling
Polishing
Graoting
Graded Rice
Nurturing the Rice Supply Chain:
A Closer Look at the Journey
4 DISTRIBUTION & ECONOMIC SIGNIFICANCE
Storage & Transportation
Storage Silos
Granaries
Storage Silos
Global & Local Transport
Distribution & Marketing
Economic Significance
Employment
Agricultural Income
Rural Development
International Trade

Challenges in Rice Production

Rice farmers are currently grappling with a multitude of obstacles that are hindering their ability to produce rice efficiently, earn a decent income, and ensure food security for their communities. The effects of climate change have been particularly devastating, with erratic weather patterns leading to droughts, floods, and extreme temperatures that disrupt the growth cycle of rice crops. This has made it increasingly difficult for farmers to predict and plan for successful harvests.

In addition to the challenges posed by climate change, water scarcity has emerged as a significant issue for rice farmers. Traditional farming methods require a substantial amount of water, but with water sources becoming increasingly limited due to growing demand, farmers are struggling to secure enough water for their crops. This has forced many farmers to rethink their irrigation strategies and explore more sustainable water management practices

Furthermore, rice farmers are also contending with pest and disease outbreaks, escalating costs of agricultural inputs, labor shortages, soil degradation, post-harvest losses, market price fluctuations, and a lack of access to modern technology. These compounding challenges have made it increasingly difficult for farmers to sustain their livelihoods and ensure a stable rice production system.

To address these pressing issues, it is crucial for governments to provide support to rice farmers in the form of improved irrigation systems, sustainable farming practices, and access to modern technology. Farmer education programs can also play a key role in equipping farmers with the knowledge and skills needed to overcome these challenges and adapt to changing conditions. By working together to implement these solutions, we can help rice farmers overcome the obstacles they face, improve their livelihoods, and safeguard food security for future generations.

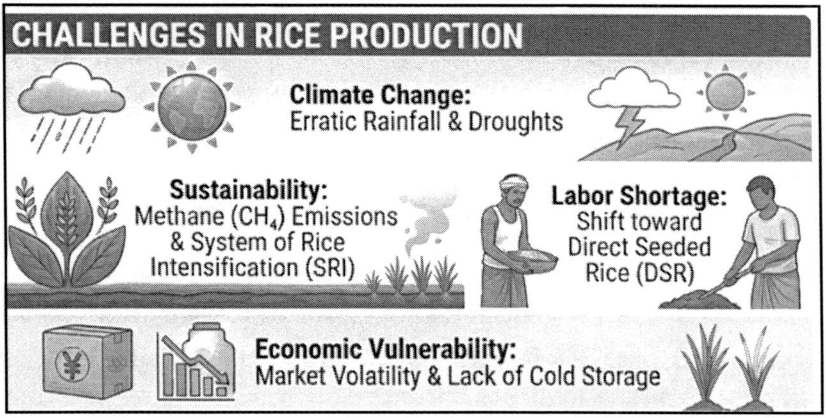

Technology and Future Scope

Mechanization covers fortyfive% transplanting, ninety% harvesting in Punjab thru 50,000 combine harvesters (₹ 20 lakh/unit). GPS-guided tractors with auto-steer level fields to two cm accuracy, saving 20% inputs.

Drones (5,000+ licensed) installation nano-urea (4 ml/liter, 50% N financial savings), multispectral NDVI mapping pest hotspots (ninety% accuracy), and variable-price spraying reducing chemicals 30-50%. Kisan e-Mitra app integrates weather, soil assessments for advisories to ten million users.

AI/ML platforms like IRRI's Leaf Color Chart app hit upon blast sickness thru phone photographs (95% precision); satellite facts from RISAT forecasts yields ±five%. Precision ag makes use of IoT soil sensors (pH, NPK actual-time) related to fertigation, boosting efficiency 25%.

Biotech: CRISPR-edited rice withstanding 3-week floods (30% acreage with the aid of 2028), drought-tolerant Nagina 22 derivatives. Direct-seeded cardio rice hybrids yield 7 heaps/ha rainfed.

Virtual marketplaces (NCRM e-trading) connect 1 million farmers to shoppers, cutting middlemen 15%. Blockchain (AgriStack) verifies GI-tagged basmati origins.

Future: Robotics like taking walks harvesters (1 ha/hour, 24/7), vertical hydroponic farms yielding 15 lots/ha in city setups, gene drives towards weeds. Swarm drones pollinate/screen one hundred ha/flight.

Area tech: Hyperspectral ISRO satellites expect outbreaks 10 days beforehand. Nano-sensors in grains detect adulteration. Policy: Rice Tech challenge goals 220 million lots by using 2035, ₹ 10,000 crore for CHCs (10,000 centers renting gadget).

Global integration: India's 20% export share grows thru weather-clever types, sustainable certifications.

Drones hovering over verdant rice paddies with AI interfaces show off futuristic farming tech.

Conclusion

Rice is far more than just a crop; it is a lifeline that sustains billions of people across the world. Throughout this study, the journey of rice—from land preparation and seed selection to harvesting, processing, and distribution—demonstrates the complexity and importance of the rice production system. Each stage of the rice cultivation process requires careful planning, labor, resources, and knowledge to ensure that high-quality grains reach consumers.

Farmers play a central role in this journey. Their dedication, traditional knowledge, and adaptability help maintain the continuous supply of rice that supports food security and rural livelihoods. At the same time, modern technologies such as improved irrigation systems, mechanized farming, drones, artificial intelligence, and biotechnology are transforming rice cultivation and making it more efficient and sustainable.

Despite its importance, rice production faces significant challenges, including climate change, water scarcity, pest outbreaks, soil degradation, and market fluctuations. These challenges highlight the urgent need for sustainable farming practices, efficient water management, improved storage infrastructure, and strong government support systems. Policies such as minimum support prices, farmer welfare schemes, and investments in agricultural research can help strengthen the rice sector.

Furthermore, the rice supply chain—from farmers and millers to traders and consumers—plays a vital role in the economic development of many countries, particularly India. It generates employment, supports rural economies, contributes to exports, and ensures that food reaches millions of households.

Looking ahead, the future of rice production will depend on the successful integration of traditional farming wisdom with modern scientific innovations. Climate-resilient rice varieties, precision agriculture, digital farming tools, and sustainable irrigation techniques will be essential for improving productivity while protecting natural resources.

In conclusion, rice truly represents the "heartbeat of humanity." It feeds populations, supports livelihoods, drives economic growth, and holds deep cultural significance. By promoting sustainable practices, technological advancement, and strong agricultural policies, the global community can ensure that rice continues to nourish future generations and remain a pillar of food security in the years to come.